JOURNEY THROUGH OUR SOLAR SYSTEM

EXPLORING MERCURY

By Robert M. Hamilton

KidHaven PUBLISHING

Published in 2018 by
KidHaven Publishing, an Imprint of Greenhaven Publishing, LLC
353 3rd Avenue
Suite 255
New York, NY 10010

Designer: Deanna Paternostro
Editor: Vanessa Oswald

Photo credits: Cover, back cover, pp. 13, 19 Vadim Sadovski/Shutterstock.com; pp. 4–5 ANDRZEJ WOJCICKI/SCIENCE PHOTO LIBRARY/Getty Images; p. 7 MAVROUDAKIS FOTIS PHOTOGRAPHY/Getty Images; p. 9 Mopic/Shutterstock.com; p. 11 Aphelleon/Shutterstock.com; pp. 15, 21 Stocktrek Images/Getty Images; p. 17 Jurik Peter/Shutterstock.com.

Library of Congress Cataloging-in-Publication Data

Names: Hamilton, Robert M., 1987- author.
Title: Exploring Mercury / Robert M. Hamilton.
Description: New York : KidHaven Publishing, [2018] | Series: Journey through our solar system | Includes index.
Identifiers: LCCN 2017003829 (print) | LCCN 2017004784 (ebook) | ISBN 9781534522848 (pbk. book) | ISBN 9781534522565 (6 pack) | ISBN 9781534522718 (library bound book) | ISBN 9781534522619 (eBook)
Subjects: LCSH: Mercury (Planet)–Juvenile literature.
Classification: LCC QB611 .H36 2018 (print) | LCC QB611 (ebook) | DDC 523.41–dc23
LC record available at https://lccn.loc.gov/2017003829
Printed in the United States of America

CPSIA compliance information: Batch #BS17KL: For further information contact Greenhaven Publishing LLC, New York, New York at 1-844-317-7404.

Please visit our website, www.greenhavenpublishing.com. For a free color catalog of all our high-quality books, call toll free 1-844-317-7404 or fax 1-844-317-7405.

CONTENTS

SMALL AND HOT

Mercury is the closest planet to the sun and the smallest planet in our **solar system**. It's made of rock and is very hot.

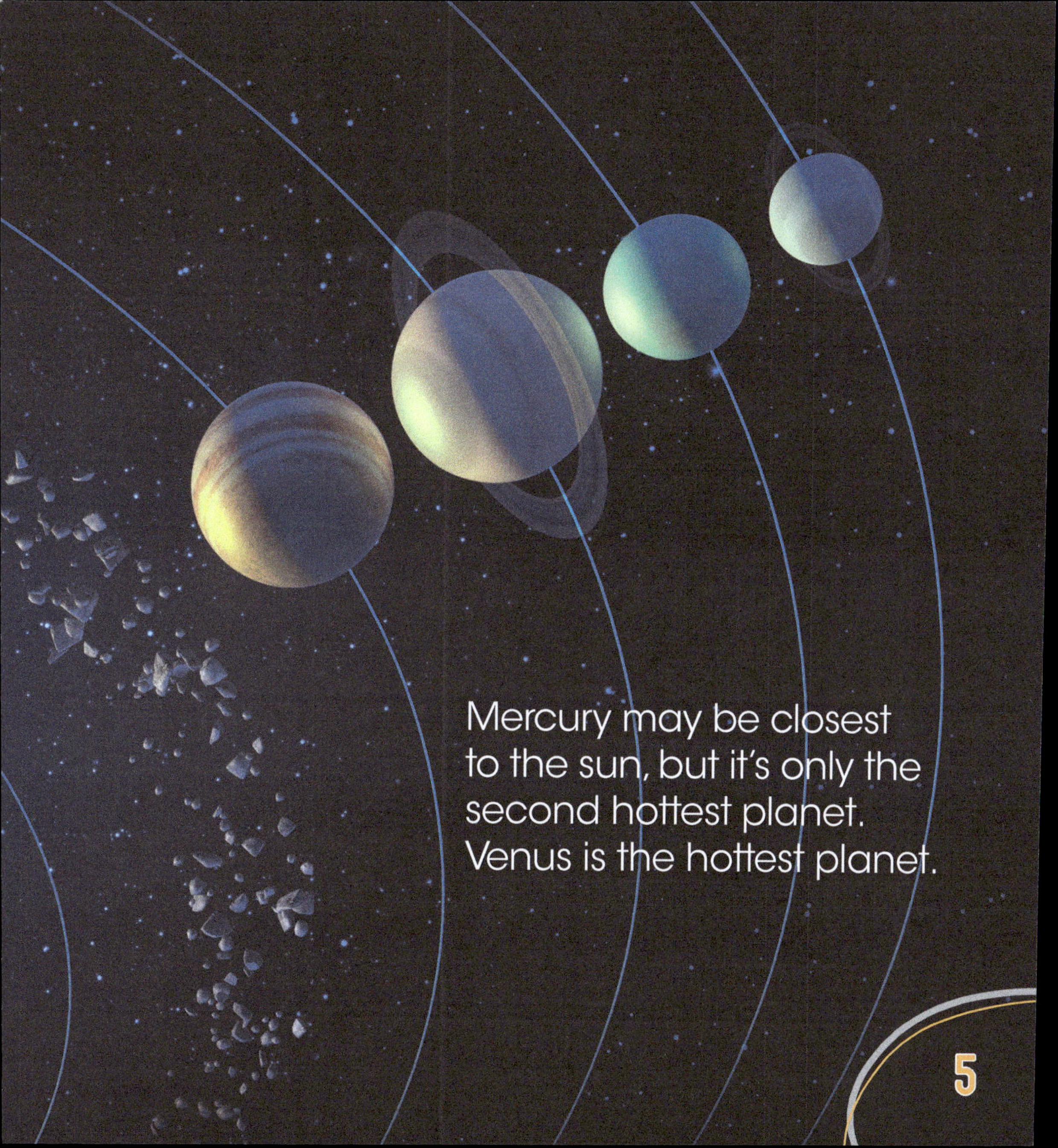

Mercury may be closest to the sun, but it's only the second hottest planet. Venus is the hottest planet.

DISCOVERING MERCURY

No one knows who discovered Mercury, but we do know people noticed the planet more than 5,000 years ago. These people called Mercury other names, such as Nabu, Apollo, and Hermes.

It's easier to spot Mercury in the sky in the early morning or just after sunset because of how close it is to the sun.

MERCURY ON THE MOVE

Mercury is the fastest planet to **orbit** the sun. The planet takes only 88 Earth days to make a full circle around the sun.

The sun looks three times larger from Mercury than it does from Earth.

Mercury is the fastest planet to orbit the sun, but it spins on its axis very slowly. It takes about 59 Earth days to spin around once.

An axis is an imaginary line down the middle of a planet.

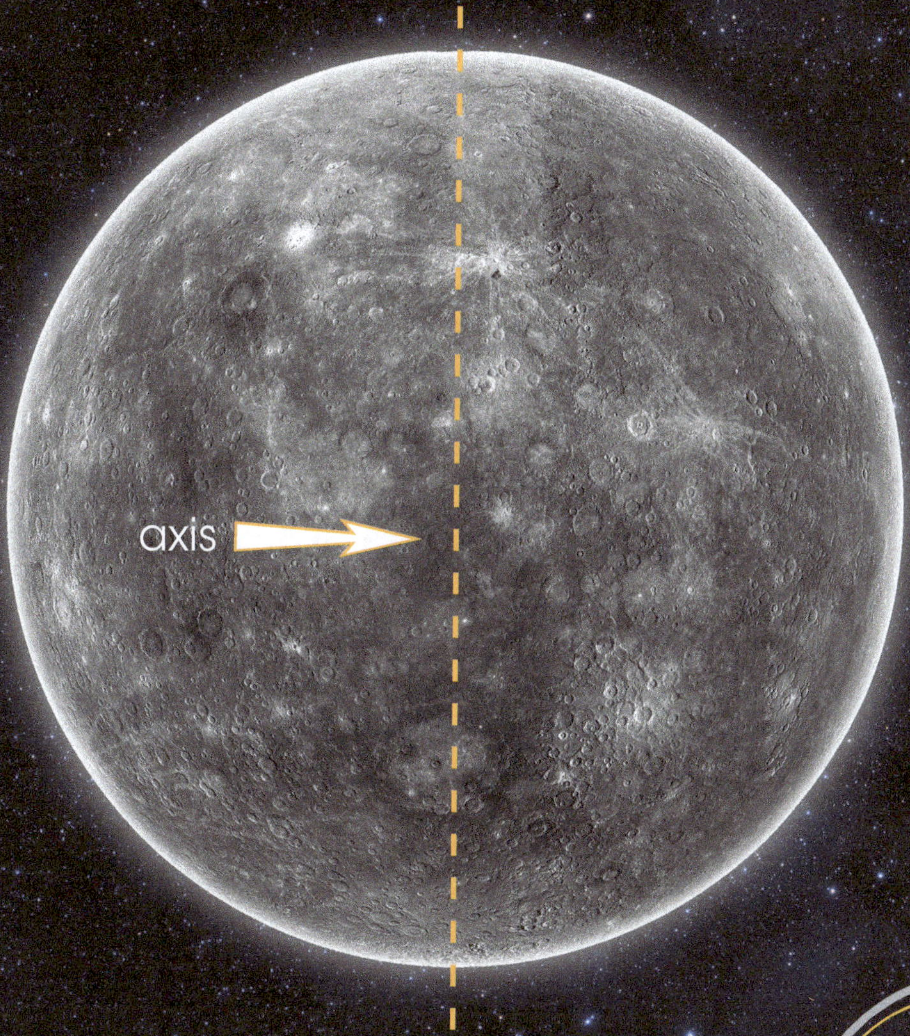

HARD TO BREATHE

There isn't much air on Mercury. People wouldn't be able to live on Mercury because there's not enough air for them to breathe!

If someone weighs 100 pounds (45 kg) on Earth, they would weigh 38 pounds (17 kg) on Mercury because Mercury has less **gravity**.

LAND ON MERCURY

Mercury's surface, or ground, is a lot like the surface of Earth's moon. It's made of mostly rock and **metal**. The dark dust covering the planet also looks a lot like the dust on the moon.

Mercury is one of the two planets in our solar system without any moons. The other planet without moons is Venus.

There are many **craters** on Mercury. These craters were made by **comets** and other space rocks crashing into the planet's surface.

The *Messenger* **spacecraft** discovered water ice in some of Mercury's craters.

LAYERS OF MERCURY

Mercury has many **layers**. The crust is the planet's outer layer. Underneath the crust is the mantle, which is made of rock. The thick center is known as the core. This layer is made of iron.

Scientists think Mercury's core is very thick.

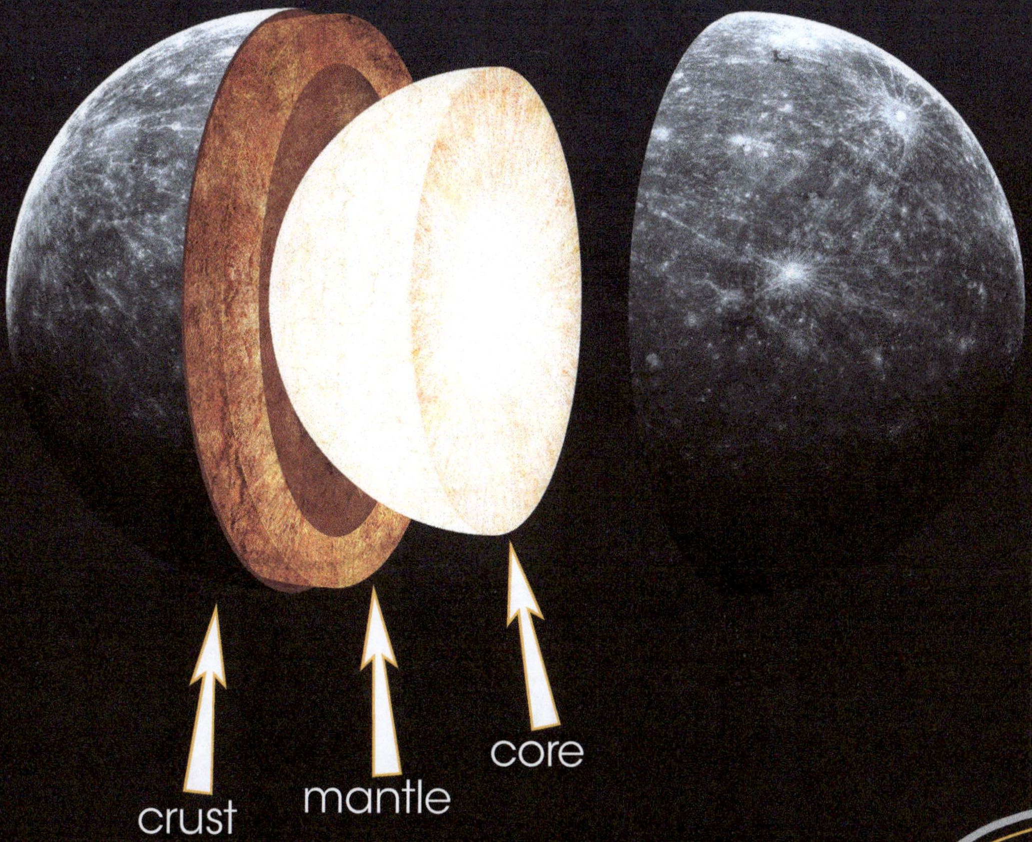

STUDYING MERCURY

Scientists have sent a few **probes** to Mercury to study the planet. The first probe the United States launched to Mercury was *Mariner 10* in 1974. *Messenger* was launched to study the planet in 2004.

Messenger studied Mercury so scientists could learn more about it.

GLOSSARY

comet: A flying chunk of icy rock in space.

crater: A large, bowl-shaped hole found on the surface of a planet or the moon.

gravity: A force that makes all things with mass move toward one another.

layer: One part of something lying over or under another.

metal: A substance found in nature that is often shiny.

orbit: To travel in a circle or oval around something.

probe: A vehicle that sends information about an object in space back to Earth.

solar system: The sun and all the space objects that orbit it, including the planets and their moons.

spacecraft: A vehicle or machine made to fly in outer space.

FOR MORE INFORMATION

Websites

NASA Space Place: All About Mercury
spaceplace.nasa.gov/all-about-mercury/en/
This website presents basic facts about and pictures of Mercury.

National Geographic Kids: Mission to Mercury
kids.nationalgeographic.com/explore/space/mission-to-mercury/#mercury-planet.jpg
Visitors to this website learn all about Mercury.

Books

Bloom, J.P. *Mercury*. Minneapolis, MN: ABDO Kids, 2015.

Owen, Ruth. *Mercury*. New York, NY: Windmill Books, 2013.

Ring, Susan. *Mercury*. New York, NY: AV2 by Weigl, 2013.

Publisher's note to educators and parents: Our editors have carefully reviewed these websites to ensure that they are suitable for students. Many websites change frequently, however, and we cannot guarantee that a site's future contents will continue to meet our high standards of quality and educational value. Be advised that students should be closely supervised whenever they access the Internet.

INDEX